Nachtflugzeug

German N-Types of WWI

A Centennial Perspective on Great War Airplanes

Jack Herris

Great War Aviation Centennial Series #3

This book is dedicated to the pioneer night flyers of the first Great War in the air.

Acknowledgements

My sincere thanks to Greg VanWyngarden for photographs and helpful feedback, to Colin Owers for photographs, Bob Pearson for color profiles and the BFW N.1 drawings, Martin Digmayer for the AEG N.I and Sablatnig N.I drawings, Jerry Boucher for the original cover painting specifically for this book, and Aaron Weaver for the cover design. I also want to thank Reinhard Zankl for his helpful information regarding production orders and serial numbers and the Deutsches Technikmuseum in Berlin for photographs. Thanks also to Mike Perkins for proof-reading. Any errors are my responsibility.

This is Update 1 of October 2013; some typographical and typesetting errors in the original have been corrected.

Cover painting by Jerry Boucher. Please see his website: **www.the-vaw.com**

For our aviation books in print and electronic format, please see our website at: **www.aeronautbooks.com**.
I am looking for photographs of the less well-known German aircraft of WWI. For questions or to help with photographs you may contact me at **jherris@verizon.net**.

ISBN: **978-1-935881-10-0**

Design and layout: Jack Herris
Cover design: Aaron Weaver
Digital photo editing: Jack Herris

Aeronaut Books

www.aeronautbooks.com

Table of Contents

Above: Becker propeller.

Above: Garuda propeller.

Above: Axial propeller.

Above: The AEG N.I generally used either Wolff propellers (above left) or Germania propellers (above right).

Introduction

Single-engine night bombers clearly illustrate one of the key differences in philosophy between the Allies and the German Air Service. The Allies tended toward quantity production of standardized designs for multiple purposes. For example, once the British F.E.2b series was in quantity production, it entered service as a battleplane, a general-purpose two-seater used for offensive fighter patrols and reconnaissance. As newer, more effective aircraft reached the front, the F.E.2b was increasingly employed as a night bomber, and eventually disappeared from daylight operations. The F.E.2b night bomber of 1918 was essentially the same airframe and engine used in 1916 for offensive patrols and reconniassance with some night-flying equipment added. The French Voisin pusher series was not quite so static in design, progressing through a number of evolving designs with more powerful engines, but it too gradually transitioned from day fighting to night bombing as aerial opposition increased.

In contrast, as tactical night bombing became more common, *Idflieg*, a strong proponent of specialized designs for different tasks, decided by late 1916 to procure specialized single-engine night bombers. Standard, general-purpose C-type armed two-seaters were used for tactical night bombing from their arrival at the front until the end of the war, but *Idflieg* wanted a design optimized for the task that could carry more bombs with a standard production engine. Drawn up in August–September 1916, the *Nachtflugzeug* (night aircraft, or N-type) specifications called for a duration of four hours, a load of six 50 kg PuW bombs totalling 300 kg, and one flexible machine gun with three drums of 200 rounds each. The required climb was 1,800 meters (5,906 ft) in 50 minutes. For navigation and bomb aiming the floor of the pilot's cockpit was to have a window, and the compass and instruments were to be illuminated for night flying.

At least six manufacturers responded to this new requirement by constructing prototypes, two of which were produced in small quantity and served operationally at the front as shown in the N-type *Frontbestand* (front-line inventory) below. The LVG N.I apparently was listed with the C.II (see p.23).

Frontbestand Inventory of N-Type Aircraft (Single-Engine Night Bombers) at the Front

	1914			1915						1916						1917						1918			
Manufacturer and Type	31 Aug	31 Oct	31 Dec	28 Feb	30 Apr	30 Jun	31 Aug	31 Oct	31 Dec	28 Feb	30 Apr	30 Jun	31 Aug	31 Oct	31 Dec	28 Feb	30 Apr	30 Jun	31 Aug	31 Oct	31 Dec	28 Feb	30 Apr	30 Jun	31 Aug
AEG N.I																				2	31	37	19	7	4
Sablatnig N.I																								2	9
Total:																				2	31	37	19	9	13

Specifications for *Nachtflugzeuge*

	AEG N.I	Albatros C.VIIIN	BFW N.I	Sablatnig N.I
Engine	150 hp Benz Bz.III	160 hp Mercedes D.III	260 hp Mercedes D.IVa	220 hp Benz Bz.IV
Wing Span (Upper)	15.24 m (50.0 ft.)	16.74 m (54.9 ft.)	15.006 m (49.2 ft.)	16.0 m (52.5 ft.)
Length	—	7.34 m (24.1 ft.)	9.38 m (30.78 ft.)	8.7 m (28.5 ft.)
Empty Weight	—	—	1,500 kg (3,307 lb.)	1,190 kg (2,624 lb.)
Loaded Weight	1,609 kg (3,547 lb.)	—	2,500 kg (5,512 lb.)	1,860 kg (4,101 lb.)
Maximum Speed	143 kmh (89 mph)	135 kmh (84 mph)	110 kmh (63 mph)	125 kmh (78 mph)
Climb to 1000m	10 min.	5 min.	—	10 min.
Climb to 2000m	23 min.	—	—	—
Climb to 4000m	50 min.	—	—	59 min.
Climb to 5000m	—	—	—	—
Armament	1 flexible machine gun, 300 kg bombs	1 flexible machine gun, 300 kg bombs	1 flexible machine gun, 744 kg bombs	1 flexible machine gun, 300 kg bombs

AEG N.I

Above: The AEG N.I was a long-span version of the AEG C.IV modified to carry more bombs, and this view clearly shows the key N.I recognition features. First, the N.I had a long wing with 3-bay bracing compared to the shorter, 2-bay wings of the C.IV. Next, the triangular bracing truss above the upper wing center section was required to strengthen the longer N.I wing against bending so it could pass the required load test. Unlike its daylight predecessor the AEG C.IV, the N.I did not have a fixed machine gun for the pilot; the observer's gun was thought sufficient for both strafing and defending the aircraft at night. The high exhaust stack directed the noise and flames away from the aircraft, which was important to maintain the pilot's night vision. Typical AEG sprayed camouflage on the upper surfaces is evident.

Like the General Electric Company in the United States, the Allgemeine Elektrizitäts-Gesellschaft was one of the more powerful and progressive technology-based companies in Germany. As part of its forward thinking, AEG established a *Flugtechnische Abteilung* (aero-technical department) at Hennigsdorf (north of Berlin) in 1910 directed by *Oberingenieur* Paul Stumpf, who pioneered the use of autogeneous-welded steel tubing for aircraft, an advanced structural technology at a time when most airplanes were made of wood. All subsequent AEG aircraft used this technology.

New assembly shops were opened in July 1915 in Nieder-Neuendorf to become the AEG Flugzeugfabrik led by *Direktor* Bassler. In the winter of 1915–1916, flight tests of an improved two-seat reconnaissance biplane, known as the AEG C.IV, were satisfactorily completed. Ordered in quantity, the C.IV was praised as a fast, rugged aircraft that stood up well during combat operations.

AEG responded to the N-type specification by modifing their successful AEG C.IV to lift a 300 kg bomb load by increasing the wingspan 2.24 meters (7.3 ft); changing the wing from a two-bay to a three-bay design. In September 1916, the AEG C.IVn prototype completed its initial flight trials, proving that it was stable and easy to fly, important criteria for night flying. AEG received a production order in December 1916 for 100 AEG C.IVn night bombers.

The first AEG C.IVn production example (C.9323/16) was dispatched to Adlershof in April 1917 for type-testing. The C.IVn wing failed repeated load tests and it was not until 7 June 1917 that sufficient bending strength was achieved by the installation of triangular reinforcing trusses over both the forward and aft center section spars. This solution, structurally efficient but aerodynamically inelegant, sufficed for a slow aircraft not expected to engage in air-to-air combat.

Timing of the initial deliveries of the AEG C.IVn and when it first reached the front are not known because it was counted among the regular AEG C.IV aircraft in the *Frontbestand*. However, by August–September 1917 the *Nachtflugzeug* (N = night aircraft) category had been established by *Idflieg*, and the first two AEG N.I bombers were recorded at the front in October 1917. The second production order for 100 AEG N.I bombers was approved in November 1917.

A reliable, efficient aircraft, the AEG N.I was primarily assigned in ones or twos to divisional two-seater and bombing units to perform short-range raids behind the enemy lines. Later in the war, some N.I biplanes served as advanced trainers.

Above: AEG C.IV 1715/17 is run up before its next mission while the aircrew and ground crew pose for a team photograph. The 2-bay C.IV wing was enlarged in span and area for more lift to carry the desired 300 kg bomb load, requiring an additional bay of struts and the inelegant bracing trusses over the wing spars. The C.IV was powered by the 160 hp Mercedes D.III, an engine in great demand for Albatros fighters, so the N.I was fitted with the more plentiful 150 hp Benz Bz.III.

Below: AEG C.IV C.1042/16 with typical AEG sprayed camouflage is probably the prototype C.IVn because C.1042/16 is a normal C.IV number. This aircraft appears to lack the characteristic over-wing bracing used on C.IVn/N.I production aircraft and retains the 160 hp Mercedes D.III from the C.IV. Three 50 kg PuW bombs are visible under the port lower wing.

Above & Facing Page: These views show the N.I's characteristic triangular bracing trusses above the upper wing center section. The trusses strengthened both wing spars to handle the increased bending forces of the longer wing. Multicolor, hexagonal camouflage similar to that used by the AEG twin-engine night bombers was applied to many N.I aircraft. Given that the N.I shared the same operational role, it is not surprising that similar camouflage was used. At right, "Franz" and "Emil" look less than enthused about their next night bombing mission. A Wolff propeller is fitted to this N.I.

Below: This rear view shows the two key AEG N.I recognition features; the long, 3-bay wing and the triangular braces for both front and rear spars above the center section. The hexagonal night camouflage typical for these aircraft is evident.

WOLFF
1320

Above: AEG N.I N.9323/16 carries air-droppable supply containers under its wings; the parachutes are wrapped around the front of the wicker containers. One of the few N.I bombers seen in unit markings, it is assigned to *FA(A)* 209. The night bomber hexagonal camouflage shows the wear from operational flying. A Wolff propeller is fitted, and the landing light in the leading edge of the upper left wing is visible.

Below: Two AEG N.I bombers are seen in in this photo of *FA(A)* 209. The aircraft on the far right in the row nearest the camera is a Rumpler C.IV; to its left in the row nearest the camera are two AEG N.I night bombers.

Above: AEG N.I N.9389/16 (possibly N.9399/16; the next to last digit is partly obscured) wears the two-color sprayed camouflage scheme and has lasted long enough for Balkan crosses to replace the original iron cross insignia. Wooden wheels are fitted due to the scarcity of rubber, and the landing light in the leading edge of the upper right wing is visible. The wood wheels indicate this machine may have been in training service when the photo was taken.

Below: Two AEG N.I bombers wearing the hexagonal camouflage scheme have collided on the ground. N.9417/16 is nearest the camera and has the letters 'AEG' in a rectangle on the lower rudder.

Above: AEG N.I 139/17, the only known photograph of an N.I from the second production series, shows the extended cabane section and the cowling panels that fold back to expose the engine for maintenance. This photograph was taken at the Bavarian training base in Lechfeld, and the aircraft is fitted with a 180 hp Argus As.III engine. National insignia is in the late style and the camouflage was sprayed on in two colors.

AEG N.I (C.IVn) Production Orders			
Order Date	**Quantity**	**Serial Numbers**	**Notes**
December 1916	100	C.9323–9422/16	Some (all?) given N designation
November 1917	100	N.110–209/17	At least one had a 180 hp Argus As.III engine

Left: Some AEG N.Is were used briefly in a civilian capacity after the war. Civil N.I aircraft *may* have been given the 200 hp Benz Bz.IV engine, which could account for the different shape of the engine cowling and over-wing radiator. The wing structure appears to be modified.

Above: A unit portrait was taken in front of this AEG N.I; unfortunately, further details are unknown. The landing lights in the leading edge of the upper wings are clearly visible.

Below: Crash of an unknown AEG C.IVn/N.I showing the night camouflage and upper-wing bracing. The night bomber camouflage is so dark it is hard to see the national insignia. The two trusses bracing the upper wing spars are prominent.

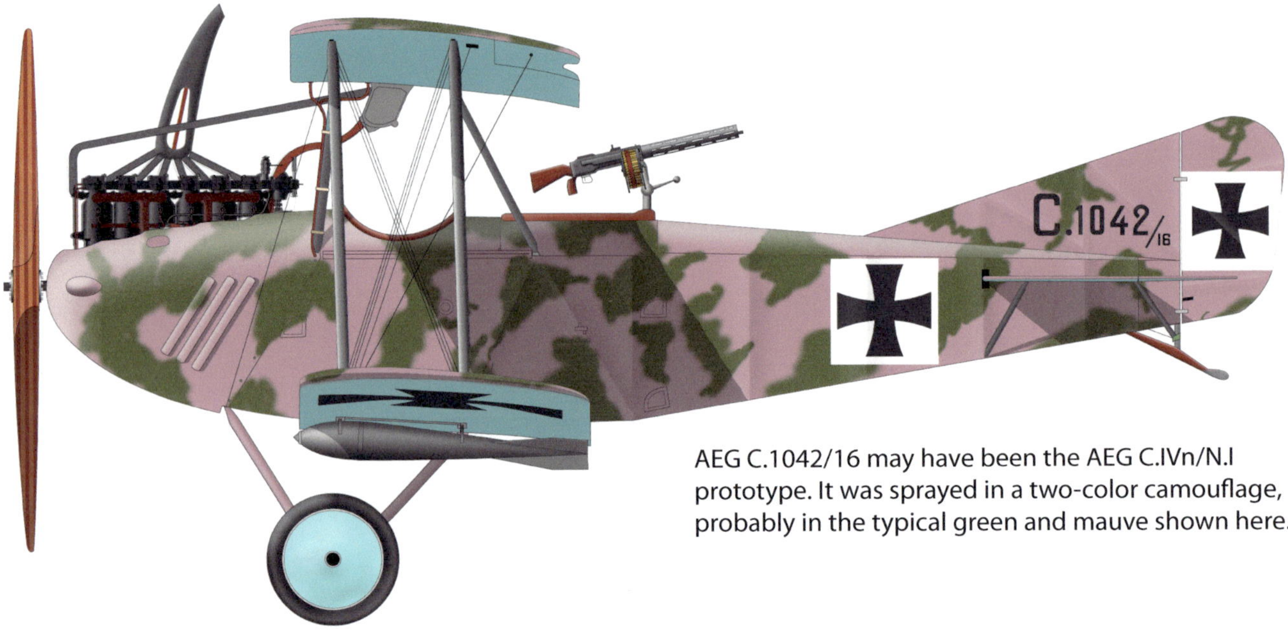

AEG C.1042/16 may have been the AEG C.IVn/N.I prototype. It was sprayed in a two-color camouflage, probably in the typical green and mauve shown here.

AEG N.I N.9417/16 is typical of many for which photographs are available. National insignia with little or no white borders are applied over the typical overall AEG dark, hexagonal night camouflage as seen on twin-engine AEG night bombers. The serial number is present in light paint and the AEG label is on the rudder below the cross.

AEG N.I N.9323/16 is virtually the only N.I for which we have a photograph showing both the serial number and a unit insignia. This aircraft was assigned to *FA(A)* 209.

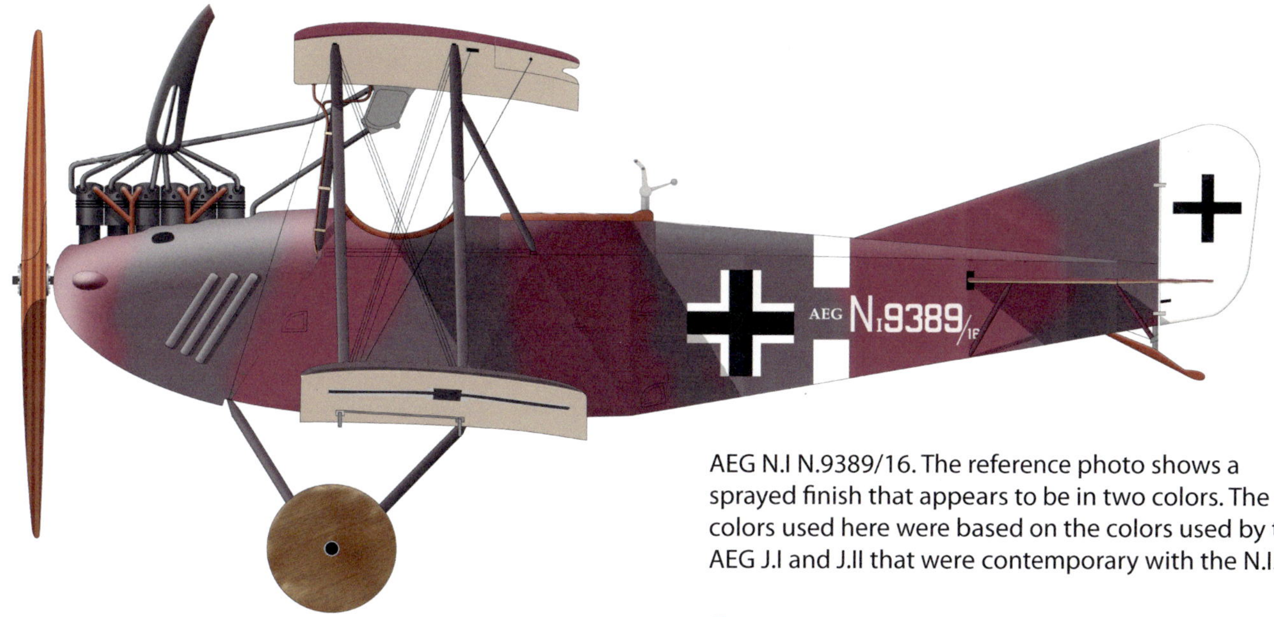

AEG N.I N.9389/16. The reference photo shows a sprayed finish that appears to be in two colors. The two colors used here were based on the colors used by the AEG J.I and J.II that were contemporary with the N.I.

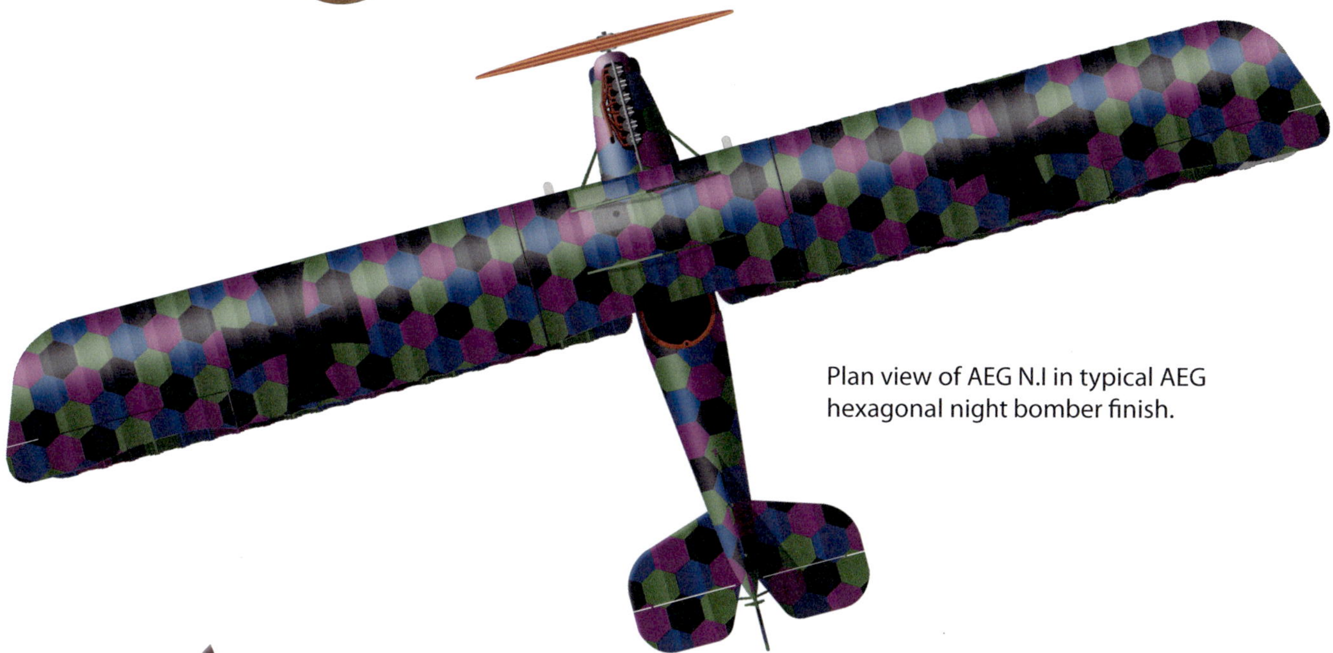

Plan view of AEG N.I in typical AEG hexagonal night bomber finish.

AEG N.I N.139/17. The reference photo shows a sprayed finish, but whether of two or three colors is uncertain. The two colors used here were based on the colors used by the contemporary AEG J.I and J.II and the pattern is representational.

AEG N.I

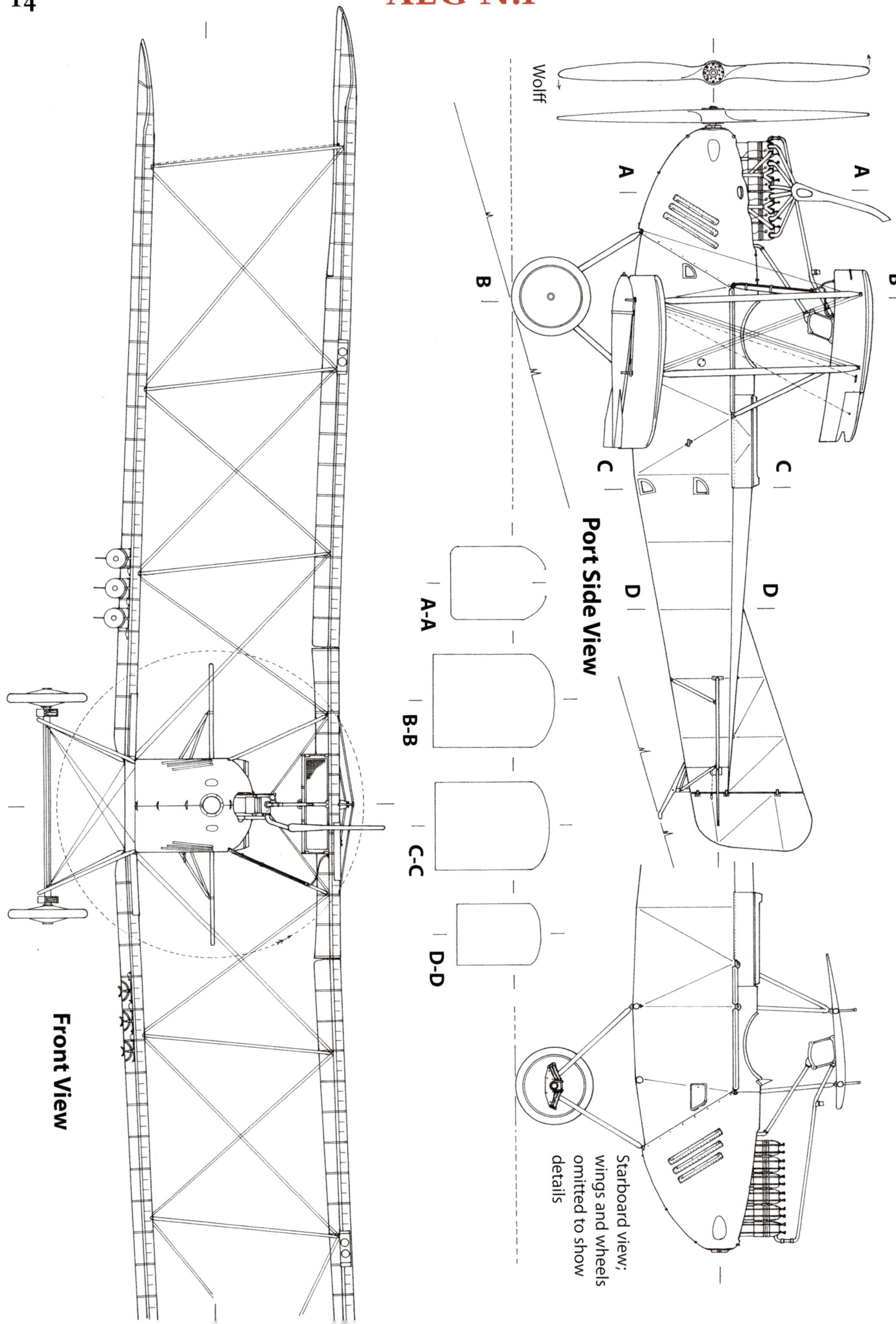
Wolff
A
B
C
D
Port Side View
A-A
B-B
C-C
D-D
Starboard view; wings and wheels omitted to show details
Front View

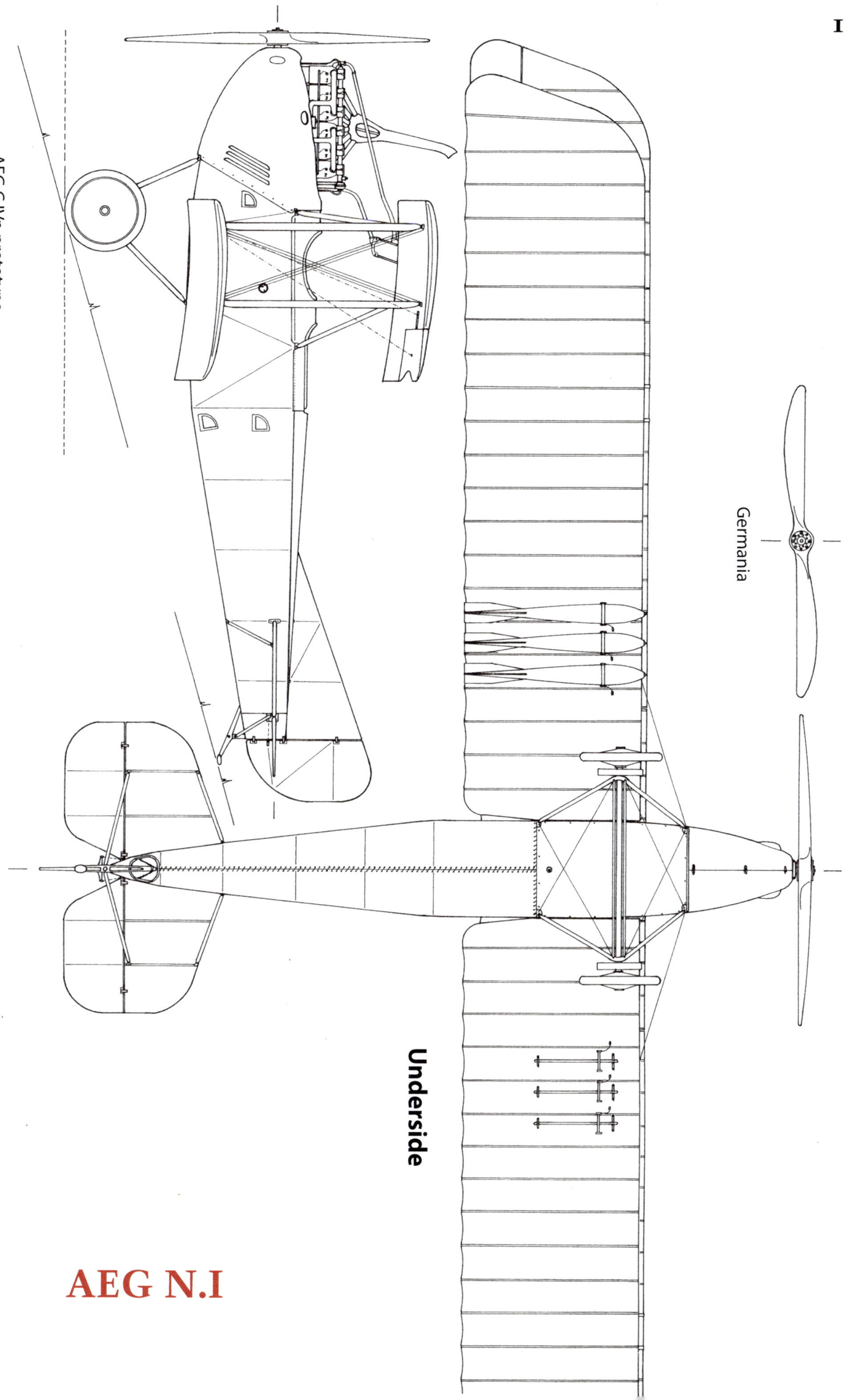
AEG C.IVn prototype
160 hp Mercedes D.III engine
Germania
Underside

AEG N.I

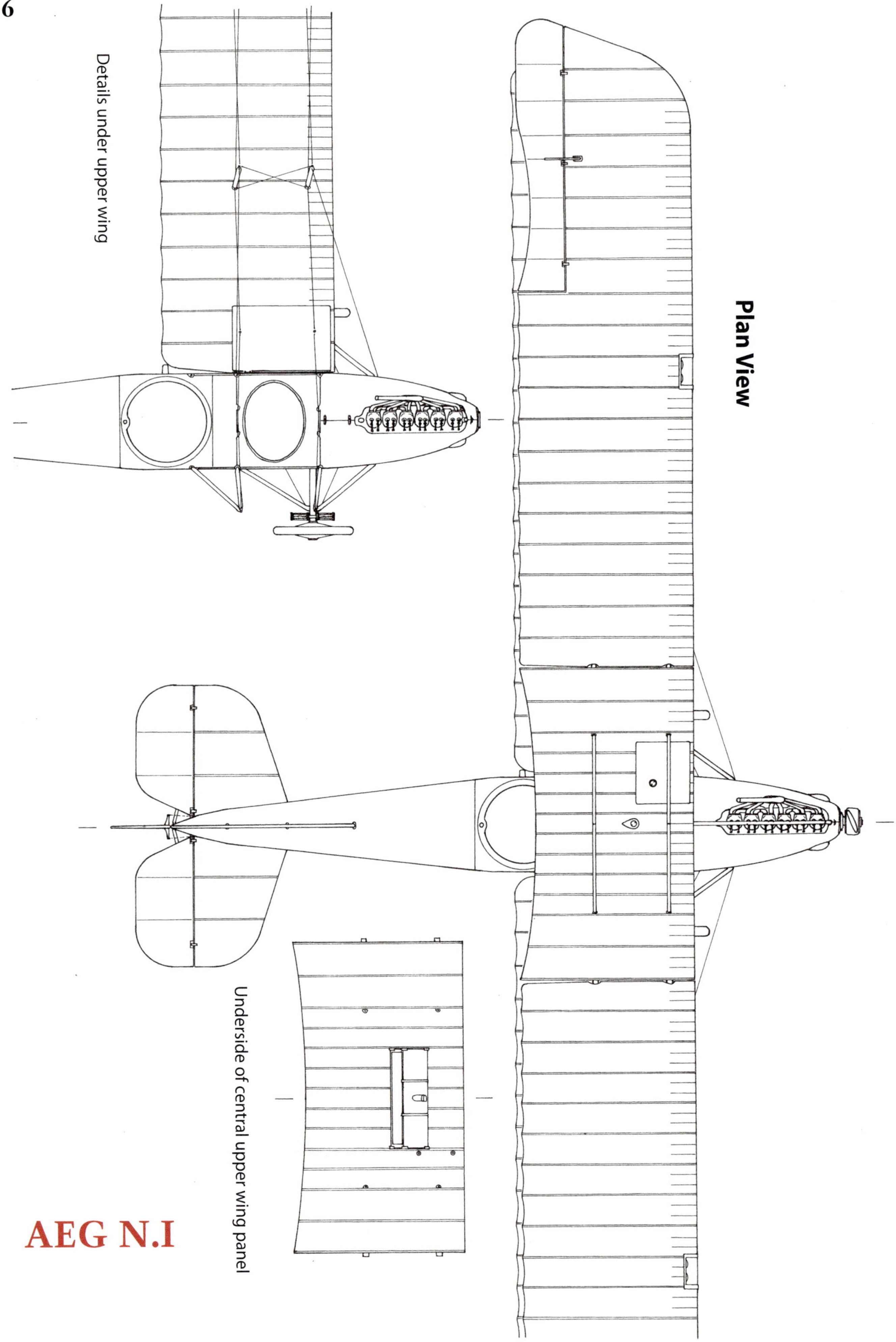
Plan View
Details under upper wing
Underside of central upper wing panel

AEG N.I

Albatros C.VIIIN

Above, Below, & Bottom: The Albatros response to the N-type requirement was the C.VIIIN, a typically-streamlined Albatros that looked much like a longer-span, 3-bay Albatros C.XII. Here it is shown carrying the required six 50 kg *PuW* bombs totalling 300 kg. However, with only 160 hp compared to the 260 hp of the C.XII, it was much too under-powered. More power might have given it competitive performance, but Albatros apparently did not experiment with that.

Albatros, the largest German airplane manufacturer, submitted the Albatros C.VIIIN for the N-type requirement. Well streamlined with great resemblence to other Albatros two-seaters, the C.VIIIN looked like a longer wing-span, 3-bay version of the Albatros C.XII. However, powered by a 160 hp Mercedes D.III engine instead of the 260 hp Mercedes D.IVa used in the C.XII, it was much too under-powered for operations, yet a more powerful engine was apparently not tested.

BFW N.I

Above & Below: The BFW N.I was a large, two-bay triplane with ailerons on the middle and upper wings connected by actuating struts. Dual wheels on each side supported its heavy weight. It was designed to carry 744 kg of bombs, nearly two and a half times the N-type requirement. It had a distinctive fin and rudder and unusual biplane tail.

In August 1917, *Idflieg* contracted with the Bayerische Flugzeug Werke (BFW) of Munich to design and build three prototype N.I night bombers. These were to be powered by a 260 hp Mercedes D.IVa engine and be capable of lifting a 500 kg bomb load. A flight duration of four hours was specified.

The dates of the prototype's completion and first flight are not known. In the fall of 1918, the triplane was fitted with test probes (venturi tubes) to explore the airflow around the airframe and determine the specific glide ratio. During these test flights two crewmen were carried to read and record the data. After the war, one BFW N.I "in repairable condition with engine" was stored at a government depot.

Above & Right: The BFW N.I was designed to carry one 300 kg bomb, two 100 kg bombs, two 50 kg bombs, and twelve 12 kg bombs for a total of 744 kg, a heavy load for a single-engine aircraft. All three wings had a 1.60 m chord. The lower wing had 3° dihedral; the other wings had none. Stagger was 6.5° and sweepback was 4°.

Left: The BFW N.I had a total wing area of 65.97 m^2, which was needed to lift its heavy bomb load. Its maximum speed of only about 110 kmh (66 mph) was so low that strong winds could have presentated significant operational problems during enroute navigation, landings and take-offs, etc. Track to the centerline of the dual wheels was 2.5 m.

The fuselage was well-streamlined with good nose entry from a spinner, but the large, flat-plate radiator beneath the top wing doubtless contributed a lot of drag. The BFW N.I appears to have been an attempt to carry a bombload nearly as large as the twin-engine night bombers being used but using only one engine. The bomb load was heavy for a single-engine airplane, and the gunner had a flexible machine gun for defense. The triplane layout was likely chosen to give the required large wing area with moderate span for better maneuverability. Few details of its performance or flying qualities have survived.

BFW N.I

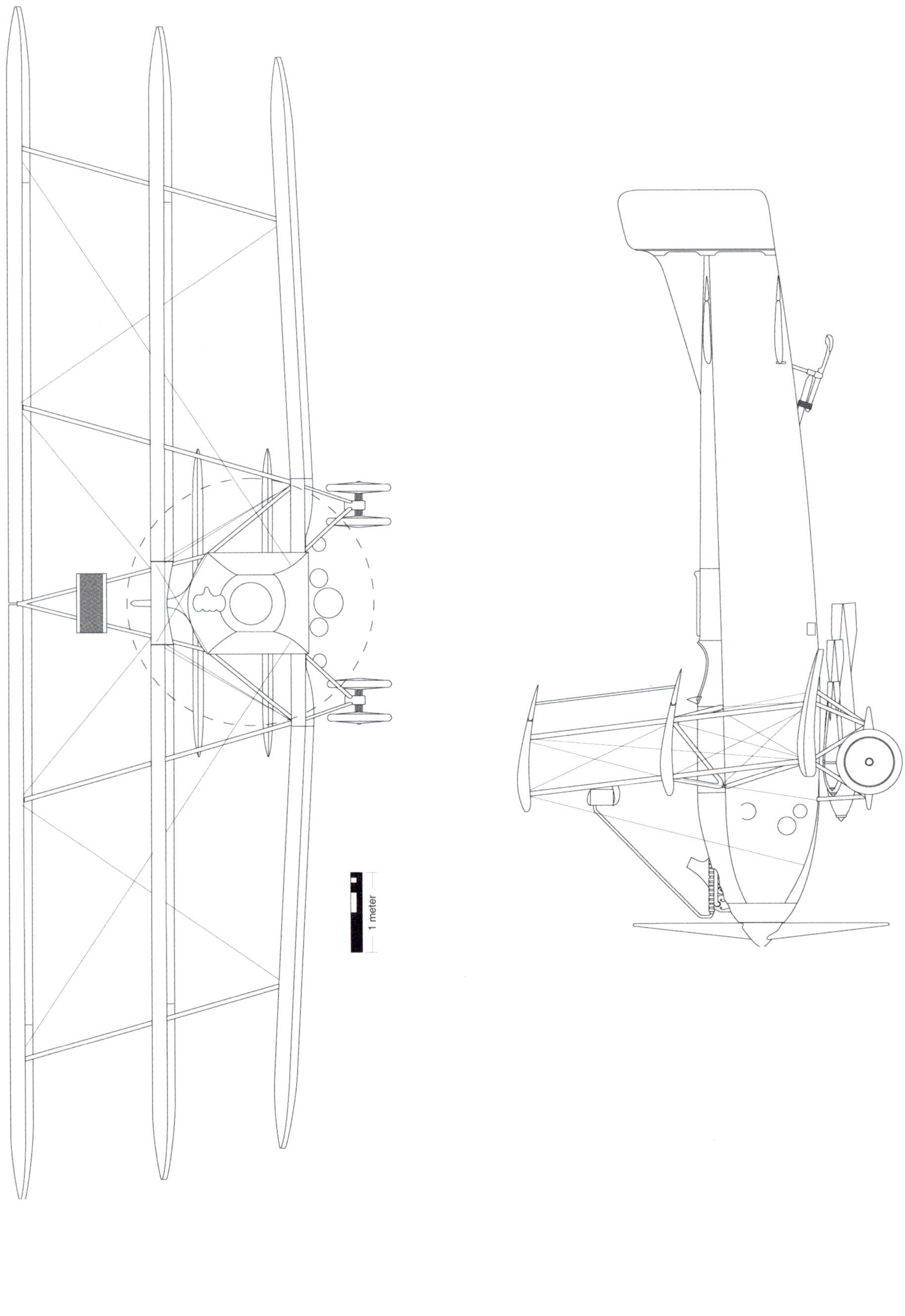

BFW N.I

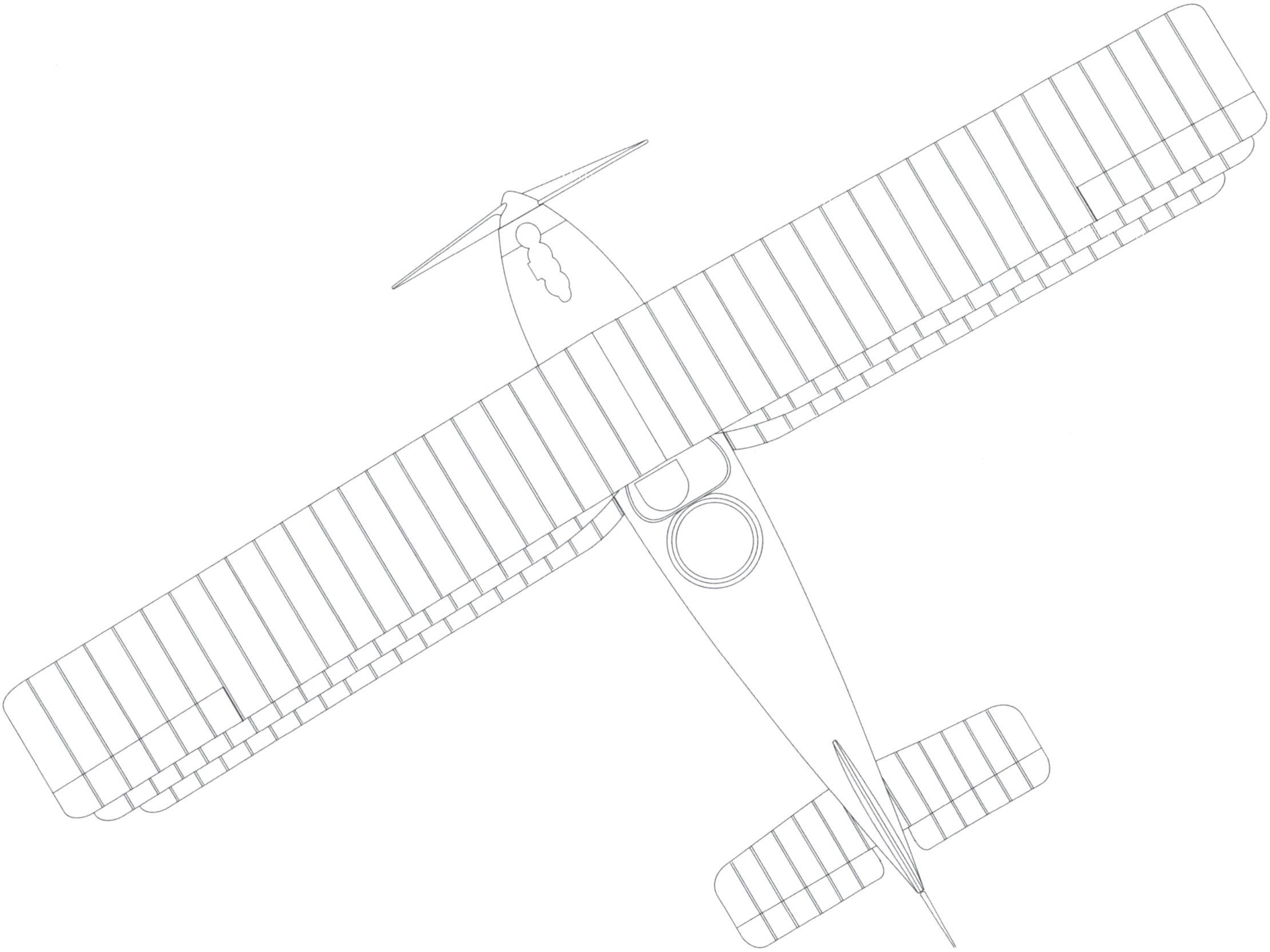

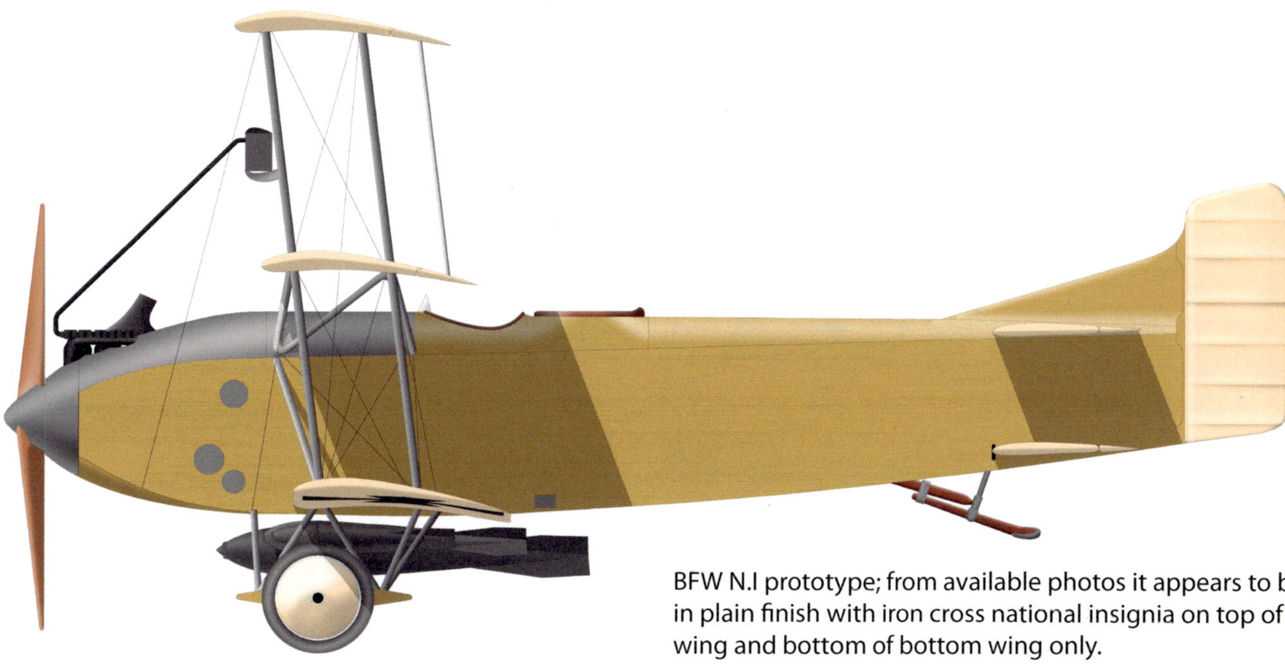

BFW N.I prototype; from available photos it appears to be in plain finish with iron cross national insignia on top of top wing and bottom of bottom wing only.

Friedrichshafen N.I

Above & Below: Designed to *Idflieg's* single-engine night bomber requirement, the Friedrichshafen N.I was another large, 3-bay biplane like the Albatros C.VIIIN. The pilot was seated well aft with an engine and long length of nose that obscured the view while landing, a feature hardly suitable for safe night operation. The required bomb load is under the wings.

Friedrichshafen specialized in floatplanes and bombers, so seemed well-positioned to develop a suitable N-type. However, powered by a 260 hp Mercedes D.IVa, it appears overly large and cumbersome with its long swept wings and very long nose that would have obscured the pilot's vision during night landings, hardly desirable for its intended night bomber role. In addition, its engine was in demand for other operational types already in production, and it remained a prototype.

LVG N.I

The LVG company was noted for a long line of two-seat combat aircraft that started with robust, unarmed B-types that led to the early LVG C.I and C.II. The B.II was a reduced-span, lighter derivative of the B.I for greater speed and maneuverability. When combat operations showed the need for armament, LVG installed more powerful engines in these two types and created the LVG C.I and C.II, respectively; both types reached the front in May–June 1915. The standard LVG C.II had a wingspan of either 12.68 or 12.85 meters; both wings were used on production aircraft and it is not known in which order they were produced or when the change took place. In any case, standard C.II aircraft were powered by either a 150 hp Benz Bz.III or a 160 hp Mercedes D.III engine and carried a bomb load of four 12.5 kg PuW bombs.

In addition to the standard C.II, there existed a version fitted with *Nachtflächen* (night wings) having a 13.80 meter span. This referred to a night bomber variant given greater wing area to lift a heavier bomb load. The post-war Inter-Allied Control Commission report mentions an LVG N.I powered by a 150 hp Benz Bz.III, which is probably the C.II with increased wing span as shown below. Unfortunately, it is not known how many of these aircraft were built nor when they served; they were certainly included in the LVG C.II *Frontbestand*.

Above: This photo of an LVG C.II probably shows the long-span *Nachtflächen* (night wings) version, also known, perhaps unofficially, as the LVG N.I. Unlike the standard C.II that carried four 12.5 kg PuW bombs, this aircraft is carrying six 50 kg PuW bombs under the wings and four 12.5 kg PuW bombs under the fuselage, a far greater bomb load. Interestingly, *Idflieg's* N-type specification required a bomb load of six 50 kg PuW bombs. Was *Idflieg's* bomb load requirement based on the long-span LVG's bomb-carrying capability as exemplified in this photo? This aircraft is fitted with a Monson propeller manufactured by the Karl Mohns & Emil Sohn Luftschraubenfabrik in Berlin. The early introduction of the LVG C.II makes it likely the LVG N.I was the first production N-type, although it was almost certainly included in the LVG C.II *Frontbestand*.

Sablatnig N.I & Related Types

Above: Sablatnig N.I 7705/17 was the first Sablatnig N.I built; it was derived from the earlier, similar C.I aircraft. As evident from the photo, the N.I was a large two-seater of conventional design and construction. The wings and horizontal tail are covered in lozenge camouflage fabric in night colors; even the rudder is dark. The large designation and serial number painted in white on the fuselage aids identification but compromises the dark, night-bomber camouflage. There is a cut-out in the left lower wing to clear the fins of a large bomb, an alternative load to the specified six 50 kg *PuW* bombs.

Sablatnig, a minor manufacturer, is best known for its series of floatplanes produced in small numbers for the German navy. Wanting to supply the German army, which bought airplanes in much greater quantity than the navy, starting in 1917 Sablatnig produced a series of prototypes for two-seat, C-type aircraft. At least five examples of the Sablatnig C.I were built, and this design was developed into the N.I.

The Sablatnig C.I was a conventional, two-bay biplane powered by a 180 hp Argus As.III and the first prototype had a streamlined propeller spinner that was omitted from the four subsequent aircraft. The C.I had landing lights in the upper wing and thus seems to have been designed for night operations from the beginning. Moreover, the C.I could carry six 50-kg booms, the bomb load *Idflieg* had specified for the single-engine night bomber type, further indicating that the C.I was designed with the *Idflieg* night-bombing requirements in mind. Ailerons were fitted to all four wings.

Developed from the C.I, the N.I had the more powerful 220 hp Benz Bz.IV and lacked a spinner, something that was clearly unnecessary for such a slow aircraft. In addition, the wing-mounted landing lights were moved to the lower wing for better

Above: Sablatnig C.I 7700/17 was the first of five Sablatnig C.I aircraft built. Landing lights were built into the upper wing for night operations and the bomb load of six 50 kg *PuW* bombs was that specified by *Idflieg* for single-engine night bombers, leading to the conclusion that it was designed to meet that specification. Interestingly for such a slow aircraft, it features a nicely-streamlined nose with propeller spinner. The large wings to lift the 300 kg bomb load are prominent, and ailerons were fitted to all four wings. Skids under the wingtips help protect them from rough night landings. The Sablatnig N.I was developed from the Sablatnig C.I through a series of modifications.

ground illumination during night landings. Like the C.I, ailerons were fitted to all four wings, and the N.I also had servo tabs mounted on the top of the upper wing ailerons to reduce the aileron control forces and improve maneuverability. Despite its substantially greater power, the N.I had similar performance to the preceding C.I.

Sablatnig N.I Orders and Production

Like most information on the N.I, orders and production details are obscure. What is known is that a batch of 50 Sablatnig C.I aircraft with 180 hp Argus As.III engines were ordered in July 1917, receiving serials C.7700/17 to C.7749/17. Technical requirements specified were: take-off and landing

Above: Sablatnig C.I 7702/17 was the third of the five C.I aircraft built. The spinner of the first prototype has been omitted, giving it a more utilitarian appearance. The servo-tabs on the ailerons are visible; these reduced the pilot's control forces to reduce his workload and improve maneuverability.

Sablatnig C.I, N.I, & C.II Specifications			
	Sablatnig C.I	**Sablatnig N.I**	**Sablatnig C.II**
Engine	180 hp Argus As.III	220 hp Benz Bz.IV	240 hp Maybach Mb.IVa
Wing Span (Upper)	16.0 m (52.5 ft.)	16.0 m (52.5 ft.)	12.5 m (41.0 ft.)
Length:	8.7 m (28.5 ft.)	8.7 m (28.5 ft.)	8.3 m (27.2 ft.)
Empty Weight	1,050 kg (2,315 lb.)	1,190 kg (2,624 lb.)	1,070 kg (2,354 lb.)
Loaded Weight	1,540 kg (3,395 lb.)	1,860 kg (4,101 lb.)	1,600 kg (3,520 lb.)
Maximum Speed	120 km/h (75 mph)	125 km/h (78 mph)	150 km/h (94 mph)
Climb to 1000m	9 min.	10 min.	4.5 min.
Climb to 2000m	20 min.	–	—
Climb to 4000m	–	59 min.	—
Climb to 5000m	–	_	30 min.
Armament	1 flexible machine gun & 300 kg bombs	1 flexible machine gun & 300 kg bombs	1 flexible machine gun & 1 fixed machine gun
Note: Sablatnig N.I specifications are repeated from page 3 for ease of comparison with Sablatnig C.I & C.II.			

run 150m, a bomb load of six 50 kg *PuWbombs*, totalling 300 kg, and a maximum speed of at least 125 km/h. These requirements were consistent with *Idflieg's* single-engine night bomber specifications, which, together with the landing lights built into the upper wing, strongly indicate the C.I was designed to those requirements. Moreover, the early AEG N.I aircraft were originally designated C.IVn; N-class serials were first assigned in September 1917, a couple of months after the Sablatnig C.I order.

It is not known when and why it was decided to build most of the aircraft as Sablantig N.I(Bz) with Benz engines. Perhaps the N.I needed more power for its night-bombing role, although the AEG N.I carried the same bomb load with only 150 hp. The 'Bz' suffix to the type designation is consistent with license production; if Benz built the airframes, it is reasonable that Benz engines would be used.

Known Sablatnig C.I serials include: 7700/17, 7702/17, 7703/18, and 7704/18 (the year '18' suffixes likely a painter's mistake). Known Sablatnig N.I serials include: 7705/17, 7730/17, 7734/17, 7736/17, 7743/17, and 7745/17. From this data it seems clear that the first 5 aircraft of the 50-aircraft order were completed as C.Is and the remaining 45 were completed as N.Is.

Above: This view of Sablatnig C.I 7702/17 emphasizes its similarity to the Sablatnig N.I. Dark camouflage has been applied for night operations.

Above: One of the last four Sablatnig C.I aircraft is ready for takeoff. The aircraft looks very much like the Sablatnig N.I.

Below: Nose and engine details of one of the last four Sablatnig C.I aircraft are shown in this damaged photograph. The landing lights are in the leading edge of the lower wings. All the C.I aircraft wore dark camouflage for night operations.

Above: Sablatnig C.I 7703/18 was the fourth of the five C.I aircraft built. The lack of a spinner on the aircraft likely made little or no performance difference compared to the first C.I prototype given how slow these aircraft were. Unlike the first C.I, the landing lights are now in the lower wings.

Below: Sablatnig C.I 7704/18, seen here without its wings, was the last of the five C.I aircraft built. Like the other C.I aircraft, dark camouflage has been applied for night operations. Did the factory painter make a mistake with the serial number? The number 7704 is part of the block of five numbers for the C.I, but the year suffix for that block is '/17', whereas the suffix here is '/18', which appears to be a mistake also made on 7703/18 above.

Above: This close-up view of Sablatnig N.I 7705/17 clearly shows the servo tab mounted above the aileron on the upper wing to reduce control forces. Ailerons are fitted to all wings with an actuating strut connecting upper and lower ailerons. The long exhaust pipe exhausts above the upper wing to avoid ruining the pilot's night vision. The cut-out in the lower left wing near the fuselage to clear the fins of a large bomb, an alternative load to the specified six 50 kg *PuW* bombs, may have been unique to this aircraft.

Below: This front view of the Sablatnig N.I 7705/17, the first Sablatnig N.I, shows the landing lights, now in the lower wings, fitted to assist night landings. The overall clean fuselage lines are somewhat spoiled by the massive radiator in front of the upper wing. Skids under the lower wing protect the structure from mediocre landings at night.

Above: A mid-production machine, Sablatnig N.I(Bz) 7730/17, shows that even the engine cowling and fuselage are covered with printed camouflage fabric in night colors. The large type designation and serial number on the fuselage are now in outline form, which is much less obvious at night, but now the rudder is painted white, compromising its night camouflage. Interestingly, the type designation and serial number are repeated on the rear fuselage at more normal size. The skid under the lower right wing to protect it against rough night landings is clearly visible.

Below: A late-production machine, Sablatnig N.I(Bz) 7745/17 reverted to the large type designation and serial number on the fuselage in solid form but retained the white rudder, further compromising its night camouflage. The normal size type designation and serial number are now moved to the fin. The skid under the lower wing for night landings is clearly visible as is the servo tab above the upper aileron.

Above: This Sablatnig N.I was used as an airliner post-war; notice the landing lights in the lower wing. The unknown pilot is at the left; Lore Birn, the woman in the center, is the wife of the man on the right, famous Austrian actor Ludwig Stössel. Being Jewish, Stössel and his wife fled to England in the late 1930s, then reached Hollywood in 1939. Among many roles he appeared in the movie *Casablanca* (as Mr. Leuchtag) and played Lou Gehrig's father in *Pride of the Yankees*.

Below: Sablatnig N.I 7713/17 is shown post-war in civil DLR colors with military insignia still applied. The post-war Sablatnig P.I and P.II passenger airplanes were derived from the N.I, but like most other German airplane manufacturers the Sablatnig company went bankrupt in the 1920s, as intended by the Allies under the terms of the Armistice.

Continued Development

Development of Sablatnig C-types continued with additional prototypes. First was the **Sablatnig C.II**; several C.II prototypes were produced that differed primarily in their inter-plane bracing. One version had two bays featuring I-struts, another version had X-struts, and the third had conventional parallel struts. All three variants had supplementary bracing struts from the upper fuselage longerons to the lower wings, eliminating the need for bracing wires on the inner bay of struts. Ailerons were fitted on all wings connected by an actuating strut. The fuselage appeared to be a semi-monocoque plywood shell similar to those commonly used by Albatros. The C.II in its various versions was more compact than the C.I with a more powerful 240–245 h.p. Maybach Mb.IVa engine. Nevertheless, speed was significantly below requirements and no production ensued.

The final known Sablatnig two-seater was the handsome, innovative **C.III** monoplane. The C.III tail and semi-monocoque plywood fuselage were derived from the C.II. Apparently the same 240–260 hp Maybach Mb.IVa engine fitted to the C.II was used, but this time a frontal radiator was employed for reduced drag. The wire-braced monoplane wing was of constant chord, and both ailerons and elevators were horn-balanced. Flying surfaces were covered in lozenge camouflage fabric. The undercarriage spreader bar was faired into a lifting surface in the manner of later Fokker designs. The C.III was well-streamlined and looked as advanced as any contemporary. Unfortunately, further details and performance of this elegant aircraft are not available, and it remained a prototype.

Below: One of several versions of the Sablatnig C.II, this was the version with conventional, parallel inter-plane struts. The exhaust has been lengthened on this aircraft to move the gases away from the crew.

In Retrospect

Like the AEG N.I, the Sablatnig N.I was assigned in ones and twos to two-seater and bombing units and labored in obscurity performing short-range, tactical night bombing. Interestingly, the Sablatnig does not appear in the *Frontbestand* inventory until well after front-line AEG N.I numbers peaked. Although this could indicate the Sablatnig had a protracted development, it is more likely that some factor related to production capacity drove the Sablatnig N.I production schedule.

The Germans preferred twin-engine night bombers for their greater reliability and bomb load, therefore the N-types were never numerous. Moreover, it is difficult to rationalize production of the Sablatnig N.I when compared to the AEG N.I which preceded it in both production and operations. Despite the significantly greater power of the Sablatnig's 220 hp Benz Bz.IV compared to the 150 hp Benz Bz.III used by the AEG, the Sablatnig N.I was slower than the AEG and apparently carried the same bomb load. While speed was not especially important in night bombing, bomb-load certainly was. And the more powerful engine used in the Sablatnig N.I used more scarce fuel to operate. Unless the Sablatnig carried more bombs, or there was some other compelling reason such as excess production capacity at Sablatnig or Benz that would otherwise have been wasted, it is difficult to understand why the Sablatnig N.I was placed in production.

The Sablatnig C.II derived from the N.I was a mediocre design deserving of its obscurity. However, the elegant C.III was a real design break-through, although the wire wing bracing appears vulnerable to combat damage and might have been replaced with stronger struts had development continued. The C.III appeared to have good development potential but arrived too late.

Above & Below: The Sablatnig C.II was a more compact, powerful development of the C.I and N.I. Powered by a 240–245 hp Maybach Mb.IVa engine, it retained the bulky, high-drag radiator in front of the upper wing and was too slow to receive a production order. It was produced in several versions; this is the version with I-struts. All versions featured ailerons on all four wings with upper and lower ailerons connected by an actuating strut and additional struts running from the upper longerons to the lower wings. The aileron servo tabs above the upper wing used by the N.I were retained by the C.II. The stubby exhaust is at the level of the crewmen's faces, clearly not pleasant for the crew. As shown below, by 1918 rubber was scarce in Germany and the tires of this prototype may have been scavenged for another aircraft.

Above & Below: Front and rear views of the Sablatnig C.II version with I-struts show more of its features. The slanting struts from the fuselage to the base of the inboard inteplane I-struts eliminated the need for bracing wires for the inboard bay.

Above & Below: One of several variations of the Sablatnig C.II, this is the version with X-struts. The bracing wires between bays appear to be connected at their intersections; this would reduce wire vibration, thereby reducing drag. The additional struts running from the upper longerons to the lower wings are clearly visible and eliminated the need for bracing wires on the inner bay. The exhaust exits above the upper wing to protect the crew.

Above: This interesting photo shows the Sablatnig C.III monoplane on the left, one of the C.II biplane prototypes in the middle, and a production N.I on the right, making for an interesting picture of Sablatnig design evolution. Although the C.II is smaller and more compact than the N.I, it retains most of its design features other than the strut configuration. On the other hand, the C.III is a major advance. Although the C.III retains the basic fuselage and tail design of the earlier types, its low-wing monoplane design had much less drag; its frontal radiator also reduced drag.

Below: This front view of the Sablatnig C.III monoplane emphasizes its clean lines and advanced design.

Below: Side view of the Sablatnig C.III monoplane emphasizes its streamlined design.

Above: This rear quarter view of the Sablatnig C.III monoplane illustrates the excellent field of fire for the gunner. The semi-monocoque plywood fuselage was well streamlined. Two handholds are visible on the lower aft fuselage for ground handling. The under-carriage spreader bar was enclosed in an airfoil like later Fokker designs. Power was the same 240–260 hp Maybach Mb.IVa engine used in the Sablatnig C.II.

Above: This view of the Sablatnig C.III monoplane shows the horn balances on ailerons and elevator and the camouflage fabric covering the wings and tail surfaces. The wings have a constant chord with cut-outs at the wing roots to enhance the observer's downward visibility. Unfortunately, nothing is known of its performance or handling qualities. On the original print metal framing can be seen on the outboard leading edges of both horizontal stabilizers that, if covered, would fair them into the elevators.

Sablatnig N.I 7705/17; the colors and pattern of this profile are conjectural and based on the available photographs.

Sablatnig N.I(Bz) 7730/17; the colors and pattern of this profile are conjectural and based on the available photographs.

Sablatnig N.I(Bz) 7745/17; the colors and pattern of this profile are conjectural and based on the available photographs.

Sablatnig N.I

Wings and wheels removed to show fuselage detail.

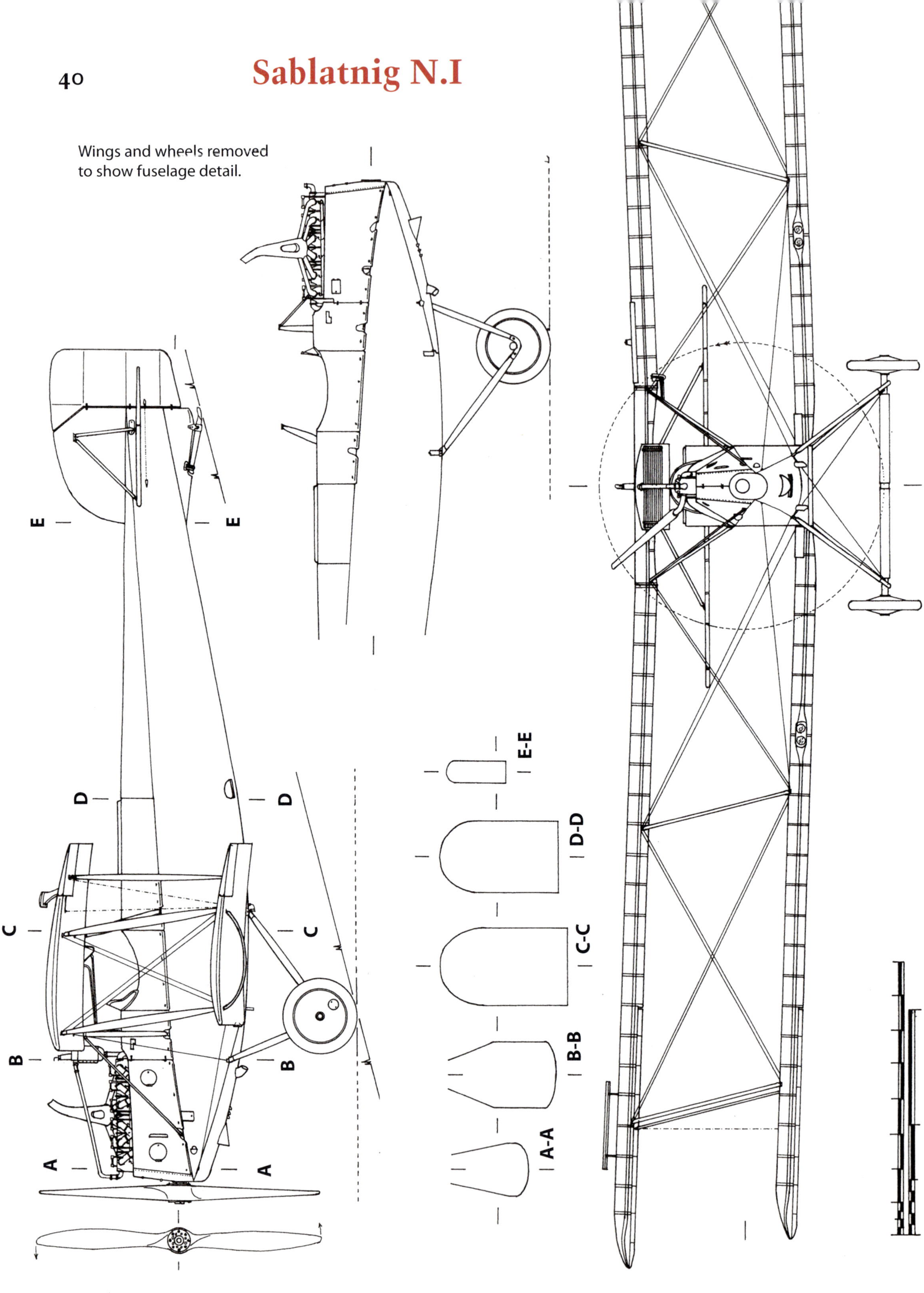

Sablatnig N.I

Plan View

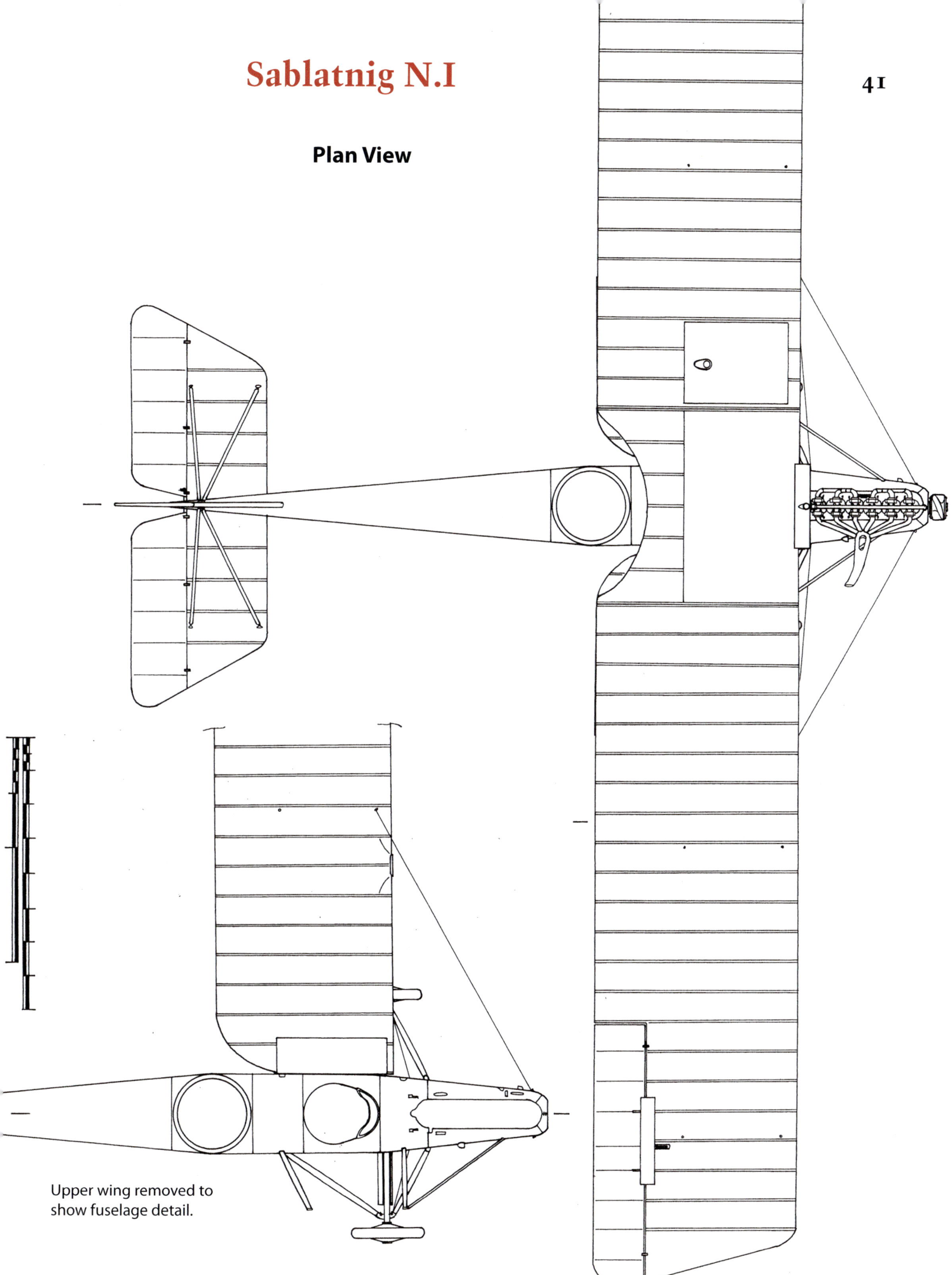

Upper wing removed to show fuselage detail.

Underside

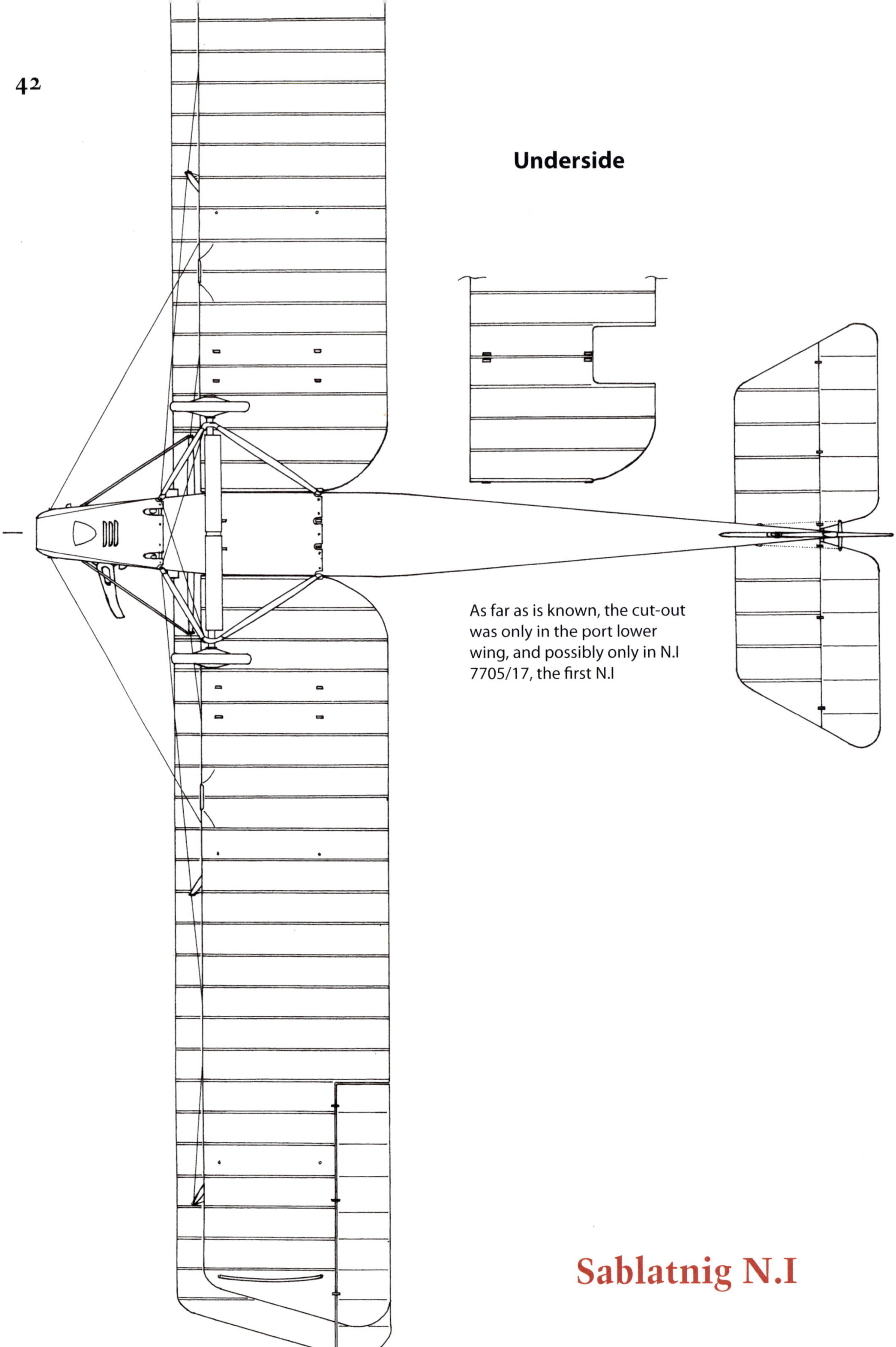

As far as is known, the cut-out was only in the port lower wing, and possibly only in N.I 7705/17, the first N.I

Sablatnig N.I

Postscript

Postwar, the Sablantig N.I was developed into the P.I passenger plane as shown in the photos on this page.

Above: Postwar the Sablatnig N.I was developed into the P.I and P.II passenger-carrying aircraft; a rear view of one of these aircraft is shown here.

Above: Propeller made by Niendorf.

Above: Propeller made by Integral.

Bibliography

Books

Gray, Peter, and Thetford, Owen, *German Aircraft of the First World War*, second revised edition, New York: Doubleday & Company, Inc., 1970.

Grosz, Peter M., *The LVG C.II*, Berkhamsted, Albatros Productions,2004.

Articles

Grosz, Peter M., "Rare Birds: The AEG N.I" *Over the Front* Vol.23 No.4, Winter 2008, p.360–367.

Herris, Jack, "Rare Birds: The Sablatnig N.I and Related Types" *Over the Front* Vol.26 No.2, Summer 2011, p.172–185.

Made in the USA
Monee, IL
12 September 2026

62975590R00029